The Best of Friends

adventures with goldens and other family members

RICHARD S. JACKSON

FITHIAN PRESS, SANTA BARBARA, 1994

To the memory of
Candy, Hunter, CC, and Copper

Printed in the United States of America

Design: Eric Larson

Published by Fithian Press
A division of Daniel and Daniel, Publishers, Inc.
Post Office Box 1525
Santa Barbara, CA 93102

LIBRARY OF CONGRESS CATALOGING-IN-PUBLICATION DATA
Jackson, Richard S.
The best of friends : adventures with goldens and other family members / Richard S. Jackson.
p. cm.
ISBN 1-56474-090-0
1. Golden retrievers—Anecdotes. 2. Dogs—Northeastern States—Anecdotes. 3. Jackson, Richard S. I. Title
SF429.G63J33 1994
636.7'52—dc20 93-45277
CIP

Contents

Part One

Candy

WHEN GRANDMA HEARD THAT WE were considering adding a dog to the family she had an answer. Within the week she advised that a sweet-natured little dachshund was already on his way via American Airlines. We protested that her generosity went beyond reasonable bounds, but we were put at ease with the intelligence that it was all part of the plan in breaking up the big house in Greenwich, Connecticut, and moving to an apartment in New York. "I think," she said, "that apartment living would be unfair to the little dear. Besides, he'd be so underfoot in our confined quarters." An impossibility, I thought, having seen the sumptuous new quarters occupying a full floor of 660 Park Avenue.

By the time of the arrival at the Louisville, Kentucky, Municipal Airport, we'd received a menu in Grandma's handwriting entitled, "Diet for the Dear Little Doggie." We also received a document from the American Kennel Club advising that a red-brown dachshund out of Fipsi von Rhein by Herman von Doerschel had been born nine months before. He apparently had not yet received a name, although it seemed a reasonable assumption that Grandma had some word for him. Further, it was noted that my wife's sister had been designated as the person to whom "this dog was disposed of." And this intelligence set up a chain of doubts. Had Sis rejected Grandma's gift? Why? Did we really want a dachshund? I had always fancied a bigger sporting dog, the kind that stretches out magnificently in front of a roaring fire, that dozes with dignity at a master's feet. Is it possible, as some people say, that thoroughbred dogs tend to be high-strung? Will he bite Mr. Pensivil, the postman, on his appointed rounds? Perhaps we'd be better advised to follow our original intention of visiting the city pound and selecting a jolly fellow of less dignified heritage. It might be considered a greater service to the animal kingdom to snatch one such derelict from the jaws of a certain death.

The excitement of retrieving the carrying case from the airline's baggage department terminated further doubts. Fipsi von Rhein's son had joined the family. I had prudently suggested to the kids that we would not free the dog until we'd reached the confines of our home. It was a good decision, and one apparently agreeable to the dog. He cowered in the back of his box, staring sullenly at his new "family." It could mean little to him that he had at last received a name, Candy. Hardly in keeping with his pedigree, it was democratically arrived at through a poll of the family, after I had unilaterally vetoed Fipsi. It was, I told my wife, too picene. He might be accused of being gay. I had learned that this was Grandma's choice, and God knows she'd had things too much her own way already. I didn't bother to pass this thought on to my wife, but it was nonetheless real.

For "a sweet-natured doggie who loves children and is housebroken to paper," Candy was contrary. At risk of limb we tried to coax him out of his traveling box. We were rewarded with snaps and snarls which discouraged further experiment. Finally, I dumped him out. He took immediate cover behind the stove.

It wasn't a joyous homecoming. For two days he ate only

when we weren't about, stealing out from his refuge behind the stove in the dark of night to gobble up what we'd prepared for his pleasure. He left evidences of his presence everywhere but on the papers that had been spread about the kitchen floor.

Loving patience and time solve most problems in life. Candy gradually accepted the family, and made a shambles of our living room carpet in the process. So much so that my wife made a detailed map of his nuisance areas against the day when we dared to take the rug for cleaning. Candy had his happiest moments trying to keep pace with daughter Bonnie. She was his favorite. I hold an indelible image of Candy rushing through the back fields, his long ears flapping to the rhythm of his short legs, vainly trying to respond to Bonnie's encouragement, as she loped along easily in search of the things that little girls and their pet dogs seek.

But he wasn't a happy dog. He made the move to Greenwich, Connecticut, with us and found it harder to adapt to the change than we did—and I was commuting to New York City, a considerable change in life-style. Although there was no dearth of peers in our neighborhood, the dogs traveling the fields in purposeful groups, Candy for the most part stayed close to home. He growled at neighbors, tradesmen, the mailman of course, and lavished his total affection on the family. By his fifth year he had developed serious trouble in his hindquarter, and he became halt. Toward the end he couldn't manipulate his hind legs, and sank into incontinence, a major sin for mature pets. He was living in misery, a victim of the structural disorder that is not unusual among his breed, forced to contend with a long, low-slung spinal column supported by short, bowed legs. By family vote, he received a quiet and dignified death at the hands of the local vet, his head cushioned in my wife's lap. We had decided that the experience would have been too traumatic for young Bonnie.

Hunter, a Replacement

YOU WOULD THINK THAT THE SAD demise of one family pet would mitigate against chancing a repetition of the experience. But this is never the case. Simple logic suggests that youthful tears can only be dried by an immediate replacement. It has nothing to do, you tell yourself, with a callous approach to the memory of the late lamented, but is rather an exercise of the theory that nothing and no one in life is irreplaceable.

Our replacement was again the result of what I preferred to call "a family consensus." Though sentiment was extremely high for a cuddly collie, my personal research on the sporting dog suggested an English setter. He was predominantly white, with the traditional black spots, and evidence of brown about the muzzle. He was extremely awkward. His head appeared to be a number of sizes too large for his body and was balanced by long, thin legs that splayed out in varying positions as he slid on a formica countertop at the pet shop. At last Daddy would have a sporting dog, docile at his feet as he read the current *New Yorker* and puffed on his Dunhill briar; and would have him despite doubts of the young and inexperienced, because I was armed with the wisdom that even a scrawny specimen such as we had in the back of the ranch wagon could sufficiently fill the void left by Candy. This setter was lively, responsive, and even humorous, as is the manner of most nine-week-old pups, and he was ours.

The skeptical could argue that his ancestry was suspect, because we never managed to procure his papers. They were promised readily and frequently by the seller and the breeder. If memory serves, it was the owner of the sire who proved to be the sticky wicket. Our setter never had, therefore, the opportunity for a family name derived from dam or sire, granddam or grandsire. We named him Hunter, although our youngest had been rather belligerent when I vetoed JJ. I felt that JJ was not only too subjective, translating into the personal initials of said youngest, but it sounded like something that might have been dreamed up in Washington, D.C., during the FDR administration. Hunter, on the other hand, was appropriate. Although it was probable that this fine sporting dog would never see a partridge take wing—we are not a family of hunters—it would nonetheless serve to remind us of the primary function of his breed.

Perhaps my wife could speak best to Hunter's formative years. I recall sharp and busy teeth. I am aware of replaced slipcovers on chair and chaise, along with a few compromised carpets. But I remember Hunter best for some of his dastardly deeds in later life, along with his pathetic fight for breath when set upon by old age.

Unlike Candy, he sought the company of the pack, as well as of any stray human being who happened upon the neighborhood, whether such intruder was a cousin, a burglar, or both. He learned early to chase cars, taught by the old dogs of the neighborhood who obviously considered this one of life's exciting pleasures. We launched a corrective campaign on several occasions. With Mama at the wheel of the car, the kids and I would hide below the open car windows. When Hunter gave chase, we'd surprise him with a pail of cold water, or a diluted stream of warm ammonia shot from a water pistol. Our aim was such that Hunter took to the game as another appealing sport, lacking a complete understanding of the seriousness of the maneuver. It was not that he was contrary. He was anything but that, as evidenced by the fact that with Bonnie as his handler he won the cup for "most improvement" in his obedience class. I've often wondered whether this experience had shortened poor Hunter's life span, because as the star pupil, his response to commands was exemplary. Yet with the award in hand, we rarely called on Hunter to stay, stand, sit, or heel. As the years multiplied his memory faltered. He knew that our occasional command required a response, but he never was sure just what he was being asked to do. As a consequence he would often lie down if requested to sit, or trot happily up to nuzzle the hand if he was told to stay.

As he gained maturity he learned that it was often profitable to stand on his hind legs and sample what my wife was preparing for dinner. So adept did he become at this dexterous act, that I once happened upon him gingerly plucking chicken livers sautéing in wine, as they simmered in the skillet. One by one he snatched them from the pan with the skill of a kleptomaniac in the five-and-dime. Not once did they appear to burn his tender taste buds.

Although he slept in a huge packing crate in an unheated garage without apparent discomfort, I often thought that he would have done much better in a traditional bed. He preferred to sleep on his back, with all four feet pointed skyward, his head over to one side. This may not have been the most relaxing posture, because he was subject to frequent dreams of possible severity, which caused him to whimper in his sleep until rudely shaken awake. But he never objected to the shattering of his dream. He had a curious smile that to the uninitiated looked as if he were ready to spring at one's throat. It was a crooked smile, where loose lips, or should I say chops, were pulled up to reveal a mouth full of teeth. If things were going right with the world and I asked him to smile for the neighbors, he was perfectly agreeable to demonstrate his dexterity of mirth, which incidentally is the only way I could be sure that it was not a gas pain, or worse yet, displeasure over my vanity in asking him to perform before company.

A Change of Venue

HE BURNED THE DOUBLE-ENDED candle of youth with the pack in Greenwich, and when six years old, moved with the family, in rich maturity, to the wide-open spaces, the hills and streams of western Massachusetts. Houses in the Berkshires are built close to the road to minimize the plowing or shoveling necessary during the snow months. This put us but twenty-eight yards or so from a major artery that, at first glance, must have seemed a paradise to a dog weaned on chasing the occasional car that traveled the quiet lane fronting our Greenwich residence. The sport had lasted but a week when a chastened Hunter limped home, an apparent victim of a confrontation with a speeding vehicle. It was a more positive cure than our poorly aimed pails of water and ammonia-filled pistols. It was only in the most spirited mood that he took to the chase after that experience. He was cured as the smoker who takes but a curious and rare puff after abstinence of some ten years or more.

He monitored that street by the hour, however, patiently awaiting our return, whether it was from a brief shopping tour, a day at the office, or a late-night arrival from an all-day festivity at the Yale Bowl. He sat under the wild cherry tree filling the role of protector of the family's worldly goods, a lonely sentinel yearning for companionship and/or sustenance. He knew our car, of course, and the sight of it was enough to bring him to his feet and get his tail going and, though he couldn't bring his floppy ears to the alert, his demeanor reflected a feeling of well-being. He never scolded if we were long overdue. Our return was enough reward.

Hunter was a creature of habit. For instance, he rarely went calling in the evening, preferring to stretch out in front of a winter fire, or under my feet of a summer evening; however I can remember a few rare instances when he disappeared in the dark. Such actions annoyed me. I would shut off all the lights and fancy, vindictively, that he'd be sorry when he returned home to a house asleep and was forced to experience the discomfort of spending the night outside. When with the morning sun he was still absent, my anger turned to anxiety. I'd hop into the car and prowl the neighborhood, fearful that I would come upon his body by the side of the road. On one such occasion I found him, along with the local pack, queued up behind a house down the street where a young bitch was being restrained in her courting by a nervous family. He came when I called him, slowly and with something of a "gee whiz" attitude when I insisted that he get in the car. Naturally, being of the male gender, I understood and I sympathized—that is, until I tried the shoe that fit on the other foot at a later date with another one of my "best friends." But that is another story.

Hunter was not a stay-at-home by any means. This would be contrary to his breed. He had friends, both dogs and humans, and he managed to keep a good dialogue with each. True, some humans he fancied more than others. But don't we all? He had favorite yards where he completed his morning toilet. Although the recipients of this largess were painfully aware of his habit, neither they nor we knew whether it was an action of endearment or derision. I prefer to believe that he was merely following the simple rule of "never foul one's own nest," and the rest was left to convenience.

I would have to say that Hunter's happiest moments were spent on the family walks. The mere suggestion of a tramp along the lakeshore, or through the raw countryside of the "lower twenty-seven" acres behind our house, sent him into a frenzy of anticipation. It was on these excursions that he spent his time, as his breed was meant to do, ranging to the right, left, and straight ahead, scaring up frightened rabbits (which he never seemed to be able to catch), mouthing sticks, overturning rocks, and enjoying the exotic scents that must have assailed his educated nose. He was always within call. Indeed, he seemed to enjoy recall on occasion. It was apparently our participation that made the walks a special treat. I assume this, of course, but I did notice that he would growl at his closest friends if they prepared to join the group. Perhaps he was saying, "Get lost, chum; this is a family party. I'll see you on Monday. Don't louse up my fun!"

As he moved into his twelfth year, he was beset by gasping spasms of shortened breath. The vet diagnosed it as a weakening heart. He spent most of his time at ease under the large elm just off the back terrace, overlooking the serenity of our lower twenty-seven. Birds, squirrels, woodchucks, fox, partridge and pheasant, and the neighbor's white cat, were now but wonders to behold, not to be chased off the family domain. The walks were less frequent and proved a tremendous challenge. Now he walked by our side, stopping on occasion to weather a fit of sneezing and loss of breath.

His last ride to the vet was the unhappiest of all. He

wasn't much on cars because it too often meant a trip toward a needle or caged confinement. This day he was so weak we had to lift him into the car. Two days later, the vet reported that he had died of a heart seizure. We didn't even get a chance to say goodbye.

He lies today, under a Vermont marble stone flat with the lawn, on the very spot that he preferred in his declining years. I often wonder if he sits under heaven's cherry tree, waiting attentively for the car to arrive, and maybe take him on a family walk.

Part Two

Golden Years

THE KIDS WERE TOLD BY MAIL OF Hunter's demise. They had matured and had left home. Such was our affection for Hunter's memory that we vetoed their suggestion that we arrange for a replacement. Would you believe it? We were overruled.

So Santa Claus delivered a five-week-old female golden retriever. As she peeked from Mama's stocking, I had a dream of glory. She was born of champion parents, her daddy being Golden Horizon's Cupid, and her mother Clifton's Fernanda. She had great-grandmoms named Feather Fetch van Lear and Lady of Cherry Lane, and then there was a great-great-granddaddy called Czar Again. Dreams of Westminster rolled softly through my head. She was a beauty, all golden except for a small white spot on the third toe of her left hind leg. But she proved to be something of a nut, perhaps taking after a great-grandmother called Beepie, or was it for Great-Grandmother Creek Koko? I had forgotten how reckless a puppy can be with household chattels. And I had never known the ridiculously vain and excitable nature of canine femininity. She didn't walk or run—she danced; backward, forward, and sideward. She was responsive and contrary at the same time, and rewarded a warm greeting by producing a wet puddle which she apparently could not restrain in her excitement. We learned never to pat her whilst standing on the Oriental rug in the hall, and denied her access to the golden carpet of the living room through the expedient of baby gates strung across the entrance. By the time we'd established some small form of order in her life and habits, the kids had started coming home for stray weekends. They applauded her playful tricks, thus encouraging her childish whims. And they came home often enough to create a disorder in her life and regimen that would, in time, test the finest obedience class.

We took a family vote to select a name, and Copper won. This sounded too masculine for such a delicate golden flower, however, so we gave her a bit of class, with a bow to her ancestry and called her Clifton's Cupid's Cupid, or just CC for the more familiar.

CC was a winter puppy. Her first real inspection of the great outdoors was a hastily shoveled path in a three-foot snowfall on the terrace. She learned in time to make her toilet in the snow. With the coming of spring, the little nut pursued her habit, relieving herself on the terrace amidst my wife's wicker furniture and potted geraniums. By the summer, I'd destroyed her natural timing by chasing her to more distant points. I was not prepared, however, for a neighbor's call of complaint.

"This is really a delicate subject," he said on the telephone, "but I can't even play baseball with my own grandkids in the backyard because of your dog."

"Chases the ball?" I asked.

"No!"

"Chases the kids?"

"No. The damn dog has left so many messes around the yard that you can't run for the fly ball."

There was a silence. I didn't know what to say. Then finally, "Gee, I don't know what I can do, but I'll try...."

"Keep the damn dog in your own yard," he rasped, ending the conversation.

We never did favor one another, and now we don't generally speak at all, but he had the last laugh. My wife had arranged for our friendly handyman to erect a fine cyclone fence around what was once the laundry yard, to contain CC in our absence. It was spacious and was fitted with a cozy house filled with cedar chips—just right for an afternoon snooze. The enclosure had a swinging door that latched on the outside. Although it deprived the family of a laundry yard, it was not an inconvenience to the lady of the house because the space had long since been commandeered by the men of the house to bed down the aluminum outboard motorboat, the fiberglass canoe, and the molded plywood single-scull shell that belonged to son number one. They had to be shunted to a haven under the giant pines to the rear of the driveway.

My neighbor could laugh because CC became a wanderer of some note, and it proved more convenient to install her in the pen than to leave her out on the town when I was at work and my wife doing her downtown errands. This changed her habits, and she reverted to despoiling her own grounds, pen, terrace, and front yard. And my neighbor's lawn lay pristine, unsullied in the Berkshire summer sun.

A Doubtful Liaison

OUR DAUGHTER, NOW MARRIED and living in Indiana, was hospitalized, and Mama left for an extended visit to give her support. This left me home alone with my TV dinners and peanut butter sandwiches. It was at this strategic moment that CC became a woman for the first time, posing problems much too mysterious for one of the male gender. Still I marveled at my wife's perspicacity. The pen proved my major bulwark. Where else could CC remain inviolate, yet relatively free, with shelter and fresh air, while I was detained at the office?

I had made inquiry of my friendly vet when evidence of CC's condition had become manifest. Lock her up, he'd said, and feed her a chlorophyll tablet with each meal.

Now, I did not entirely trust the handyman's enclosure because on one side the wire lay almost against a stone wall that retained the hill from which the laundry yard had been scooped. The fencing was the same height as the retaining wall, separated by a gap of approximately five or six feet. This should challenge even the most energetic male, but I had heard of a macho dog's persistence in such cases.

The tablets performed beyond my wildest expectation. To almost the man, or should I say the dog, the males of the neighborhood never seemed to get the scent. All except Tinklepaugh, that is. I don't really know what his proper name is, but he is the irascible pet of Herman

Tinklepaugh, the baseball magnate who lived next door, one of those guys who argued about the common boundary of properties, borrowed our gardener working on Mama's prize flower bed to "just finish up a little job that needs a professional hand," and in general contested every decision that existed in the neighborhood. He was short, cantankerous with age, bowlegged and sly. He had amassed a veritable fortune from wheeling and dealing in real estate. Further, although seventy years of age, he had a persistent record of attracting young and beautiful girls, which made him the envy of all gentlemen who were younger and more handsome. His dog was a carbon copy in dogdom: sullen, short, and bowlegged. He was black with a splash of white on the tip of his tail, had bloodshot yellow eyes reminiscent of his master, and he was just as sly.

We'd managed to live at peace with the Tinklepaughs, as long as they stayed on their side of the boundary. This was the unwritten law, and CC helped to keep it enforced. But the summer of CC's heat, dog Tinklepaugh was not fooled by the chlorophyll tablets, nor by CC's constant confinement. They were nose to nose through the cyclone fence when he guessed that I was not about, and apparently readily accepted by his erstwhile enemy in her fertile condition. But what he knew he kept to himself. That handsome shepherd down the street, the Irish setter, even the big-chested golden with the rampant feathers that floated in the breeze as he ranged the neighborhood, none of them knew. And Tinklepaugh had no plans for telling.

It was the ninth day, some half-hour after I'd fed CC her breakfast and repaired to the kitchen to eat my own bowl of Wheaties, that Tinklepaugh made his move. When I passed the pen, natty in pin-striped business suit with leather briefcase swinging at my side, I noted with horror that the heavy wire door hung awkwardly from but one of its hinges. CC was gone. The citadel had been breached! Terror tore at my heart. Tinklepaugh, I knew, was hardly a fitting mate for my pretty. She deserved much better. At least a "best of breed," if not "of show." Her mate should be a handsome male of her own persuasion, with the impeccable credentials that go with all weddings of the highborn.

I dropped the tools of my trade and cast about for the missing. Aha! There she was, stretched out on the grass at the edge of the driveway, watching my frantic reactions. She looked relaxed and, perish the thought, sated. Lurking under the canoe amidst the giant pines I saw Tinklepaugh, a surly study of satisfaction. He turned when our eyes met—my soft, intelligent browns and his bloodshot yellows. He gave a supercilious grunt, dug his hind feet into the soft turf, and sprayed dirt and pine needles in my direction as he sauntered off for the distant boundary.

It is too early at this writing to recount the fruits of this dastardly deed. That is a tale for another day. CC's sins have earned her a protracted stay at our friendly vet's emporium. The hinges of her open door have been altered, and the dirt threshold under which she started to dig her escape has been renewed with a sturdy rock laid by my own hands. I have procured a fresh supply of BBs and hunted up my son's favorite shooting iron of a youth gone by. I promise that I shall not aim for Tinklepaugh's yellow eyes, because even enduring hatred disallows such barbarity. But I'm not beyond hitting him in the other end.

We are left with but one hope, as our misery measures the time of decision. If it is to be, I hope that they will all be males. I love my wife and daughter, and, yes, I love CC; but three women in any family can strain a man's patience.

A Proper Marriage

CC HAD AN IMPECCABLE PEDIGREE. Her selected husband, Clarence, was formally named Champion Clifton's Caspar of Round Hill. He was a sire of note in similar philanderings, and one who had earned the right to have the calling of champion. Clarence was a large fellow, obviously handsome and well-mannered, but with a sufficient spark to otherwise quiet eyes that suggested an undertow of fun and fancy. There was little question that he would relish a good chase for the thrown ball or stick, and this would make him one with CC, whose pleasures in life could be simply defined as a continual chase for any thrown object, perhaps sustained by food, water, and an occasional loving pat of acknowledgment by friendly hands.

Still, being technically in heat, it was presumed that her horizons had been somewhat expanded, although she gave little indication of this as we arrived at the site of her scheduled tryst. After enduring a hasty sniff from Clarence and some of his live-in harem who skittered about the grounds, she happily joined in a game of ball that was taking place on the back lawn. But the purpose of our visit was not gaming. CC had been subjected to what we suspected might have been a precarious union with a low-born neighbor some six months previous. Fortunately, the fruits of this liaison had never matured. Tinklepaugh was the name we used for this neighbor, a fellow of mixed heritage, of bowed legs, evil yellow eyes, and dull black coat with a white spot on the tip of his tail. He was hardly fit for our lovely highborn golden. So it was with some impatience that we sought this union for CC with one of her kind before chance and desire consumed our lady and we should all have to adjust to its consequence.

CC and Clarence were led to the honeymoon area, and all but the handler (and owner of Clarence) were banished from the adjoining chamber. We were told that CC had "flagged" upon being segregated with her mate, flagging being a raising of the tail as an indication that she would fancy the intimate attentions of her intended. Once the deed was done, the lovers reappeared in a rather compromising position. They were, in the terminology of the breeders, "tied," still joined as husband and wife, and facing, curiously, in opposite directions—or almost so. This is a precaution taken by nature to insure the consummation of the union, the better to permit the passage of sperm toward fertilization. It is also the very posture in which neighborhood dogs are often discovered and prematurely separated by being doused with a bucket of water, or worse, by a broom-handle or stick. The laying on of such tools could hardly equal the pain that certainly the male, and perhaps the female, would experience through the abrupt disengagement that nature sought to encourage. But CC and Clarence, for the moment as tied as Siamese twins, shuffled toward the edge of the pen in which they were enclosed and watched with obvious interest and envy as the ball game continued on the adjoining lawn. Passion and its compulsive desire were well behind them. What lay ahead was the thrill of the game on the lawn, a game they joined in God's own time.

It is difficult to plan around birth or death, and so it was that we were on the ski slopes of Aspen, Colorado, at CC's time of need. She had been provided for, however, and was temporarily residing at the veterinarian hospital, which, as good planning goes, was exactly where she should have been as her time approached. Our first indication of the blessed events was a telegram from our vet, read to us by an incredulous telephone operator. It read, "CC had seven healthy sons. Mother doing fine."

The operator said, "My gawd, that must be a new record!" Hardly; but then, I suppose seven sons and no daughters is something of a rarity even in the canine world.

Family Life

THE MOTHER AND HER SONS CAME home when we did, four days after the brood had been introduced to their new world. They were quartered in a large whelping box with a swinging door to provide CC with an opportunity to leave at will. Around the inner periphery of the box we'd inserted round bars, about an inch above the floor and a like distance from each side. This, we had been advised, would keep the young pups from being inadvertently smothered.

They were a delightful family. Seven round balls of fur, their heads appearing to be screwed onto bodies by wrinkles of flesh that bulged at the shoulders. Noses were wet, black, shining, and pugged. Ears hung almost to the ground, and they had yet to open their eyes, although this hardly interfered with finding an available food spigot when the spirit moved them, which was often. CC displayed the uncanny finesse of a mother. True, she'd been faulted by the midwife for severing umbilical cords too close to her offsprings' tummies, but this they say is not unusual for a rookie mother who could hardly know when her own production line would come to a close. It must have been an exciting time for our CC.

She was an attentive mother. One could almost see the pride in her demeanor as we fawned over her family, handling this little fellow and then that one. It having been a spring birth, we were blessed with warming weather, and in good time we introduced the pups to the great and wide outdoors. By this time the pups had managed to keep their eyes open. We had started the process of trying to identify one from another. Their temporary names were descriptive for obvious reasons: White-on-the-Shoulder, and White-on-the-Chest, Big Blonde and Little Blonde, Lazy, Sleepy, and No-Name.

As they graduated to solid food, each had a small dish of his own. Although they started each meal neatly lined up with one mouth per dish, it soon became evident that a brother's meal had more appeal. There was a good deal of switching. Some were more greedy than others, or had bigger appetites. To the stronger go the spoils, but we made every effort to see that each pup got his share.

About the third week of their new life we had a scare. By now they could clamber over the three-inch threshold of the door to their box, and they did so whenever the door was swung open. It meant that it was time for some kind of excitement, the lead dog generally assuming that he'd better check out the kitchen first. On this gray morning, they tumbled out of their box and, rather than making a dash for the food, they created a most confusing tableau. They had thrown themselves on their backs, and kicked legs in the air or squatted on their stomachs and licked their paws. At the same time, all were putting up a horrendous racket. To the dog, they were crying and whining. I noted that they had lost the sparkling sheen of their coats and looked gray and bedraggled. They were obviously ill.

We called the vet. What can it be, all seven ailing at once? He suspected that it was something that CC had eaten and passed along through her nursing.

"Check the box, does the paper look unduly wet?" We

did, and it did. The mystery was solved. Something in her diet had caused excessive urinating on the part of the family, and they were literally swimming in their acid output, which burned their paws, dulled their coats, and caused discomfort. I can't precisely recall the remedy, but it was my impression that nature was allowed to run its course. By lunch time things were back to normal.

By the fifth week they were essentially weaned, though reluctant to give up the handy spigots. CC would take them down into the back field and show them how to retrieve sticks, balls, or pine cones that she insisted we throw. It was not unusual to have as many as five tiny mouths affixed to one long branch, with CC gently guiding the pack back to the terrace for the next throw.

My wife had a notation in her daybook indicating that it was time to remind the vet to remove CC's stitches, which had been used to close out her birthing. A telephone call was placed to make arrangements. But CC, said the doctor, had already taken matters into her own paws. The vet lives but a meadow and pine grove away to the rear of our house, about a mile from his hospital. CC had, that very morning, appeared at his front doorsteps unannounced, but apparently expectantly. The vet took the hint and removed her stitches on his back porch. He'd meant to call and advise us that the deed had been done.

Above: Candy learned to love his adopted family.
Below: Mama takes a winter walk on Lake Onota with Hunter.

Opposite:
Above: CC, as a five-week-old pup, learns about the Berkshire winters.
Center: Proud mother and sons.
Below: CC teaches the art of retrieving sticks.

This page:
Above: Each pup had his own dish.
Right: KP duty after a meal.
Below: Mother and son in repose.

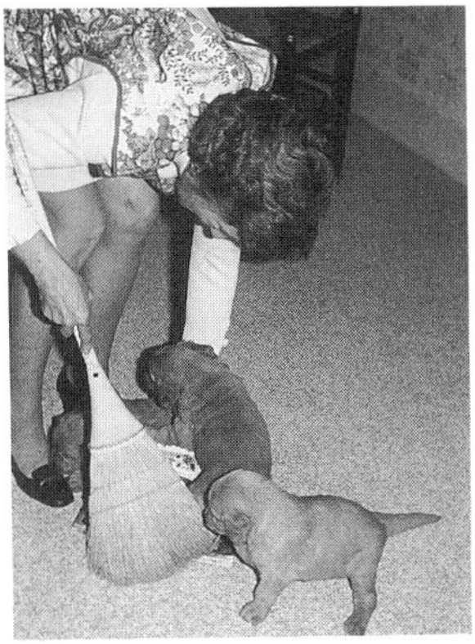

Above: CC plunged off docks with abandon,
while Copper watched in amazement.
Center: Copper loved the family to distraction.
Below: How the dogs loved it in Maine,
with the entire ocean as a playground.

Part Three

Copper of West Street

THE TIME HAD COME TO SEPARATE the family of seven male puppies. We'd put an ad in the local press, and also by word of mouth friends and interested buyers appeared to inspect the brood. Two we sold to friends, another to a south county farmer who needed a workmate. One was privileged to be a wedding present for an upcoming groom, a gift of the bride-to-be. One was returned to Clifton Kennels to satisfy the stud fee, and two we retained within our family. White-on-the-Chest was a special Easter gift to our daughter, and he became known as CC's Golden Nugget. That left White-on-the-Shoulder, the runt of the litter. He was a spindly fellow with a darker coat than his brothers, and, as he matured, had the look almost of an Irish setter. Smaller than his brothers, he grew tall, but always remained thin. And because of his coloring it was an easy matter to find a proper name for him. He became Copper of West Street.

He was at once aggressive and inquisitive, impatient, selfish, awkward, loving, and loyal almost to a fault. It was always Copper who had demanded his food first of the brood, and was willing to bark for it before his brothers realized that this was a useful ploy. Indeed, it was also Copper who was the last to despair of suckling his mother for one last time. He had an oriental look, slanted eyes, a large head that he barely outgrew with age, and large legs with enormous feet, legs that were a sharp contrast to his mother's shapely feminine limbs and delicate feet.

If CC's greatest pleasure in life was chasing a ball, Copper's was just to be around the family, and hopefully to be on the receiving end of affectionate pats or endearing comments; in brief, to be the center of attention. From an early age his happiest moments were experienced curled up in my wife's lap, although we had to discourage this little habit as he put on weight. But he never fully outgrew the desire, and he arrived at what he must have thought a workable compromise. He would haul his upper body into one's lap, but one hind leg remained touching the floor. This he assumed met our requirements.

Copper never knew what it was like to exist without his mother, and he never fully lost his dependence upon her. He grew to be much taller than CC, and he took over her role as watchdog for the homestead. His bark held a throaty warning, diminished in intensity by a simultaneous wagging of the tail. With the first sign of friendship from any interloper, Copper would gladly give away the family jewels, settling for a friendly pat. He would, I feel sure, have licked the hand of the most vicious intruder.

There was one exception to this that I can remember. My wife returned home from shopping one sunny morning and found the front hall desk in a mess, drawers pulled out, the contents of some emptied on the floor. Naturally her first thought was of her slovenly husband, the undoubted perpetrator of the deed. But as she moved through the hall she sensed the presence of an intruder. This prompted her to pass quickly out the terrace door, where she threw open the gate of the dogs' pen, and out CC and Copper popped. Perhaps it was the chemistry of her growing fear that excited the dogs, but they and my wife proceeded back to the house on the run prepared to do battle, the dogs in full cry, the hair on their backs uncharacteristically bristling, their fangs bared. The uninvited stranger had read the signs and rushed out the front door as the menacing trio appeared from the terrace. The chase was brief. The intruder made it to a parked car with a henchman at the wheel, and sped off just a few paces ahead of the pursuing goldens, with my wife some thirty yards to the rear. Like Hansel and Gretel, the thief had dropped unopened packages of cigarettes as he ran, the only booty he had apparently managed to collect before his flight. He was never apprehended, nor was he identified, but for that matter, he has yet to return to complete his mission.

Mother and Son

IT WAS NATURAL THAT COPPER AND CC would travel together. Generally they had a neighborhood walk in the morning and stopped in at the kitchen door of each house along West Street to see what they could scrounge. Some of their stops were hardly rewarding, because some folks just can't abide dogs, benign or otherwise. Some were quite the opposite. The Browns down the street looked forward to the morning call. They had become friends. Mr. Brown told me "it took a while to gain the confidence of the female, but the male was very friendly, and accepted a cookie as first offered." Of course, that was one certain way to win Copper's affection. He had a hearty appetite for one so thin, but I figured being of such an excitable nature, he burned more energy than the average dog his size. Copper never really ate his meals, he inhaled them. When he was growing up he used to finish his supper well in advance of CC and then he'd circle around her while she ate until a motherly growl would send him on his way.

The relationship of the two dogs was something of a paradox. I have often heard it said that familial ties are not honored among dogs. Once "out of the nest," it is every dog for himself. And this would appear to be true—if, that is, dogs could be considered not to have any standards with regard to incest. Certainly CC's Golden Nugget, who lived quite apart from our dogs but who did on occasion pay us a visit, viewed his mother as very fair game for his masculine attentions, despite the fact that we had taken the precaution of having CC spayed after her satisfactory family experience. Even Copper would on occasion be roused by masculine inclinations, though for some reason poor Copper never really seemed to get the hang of what he was supposed to do. I recall his one great opportunity to perpetuate his lineage—a license to spend the entire night with a perfectly stunning lady of his own breed, a gorgeous golden who belonged to the local bank president. We delivered Copper to the kennels where the marriage was to be consummated, and on the way I suggested to him what was expected. I thought we had quite agreed that it was the chance of a lifetime. I was a bit concerned when Copper looked mournfully at my departure rather than expectantly at the bride to whom he'd just been introduced. Yet I was confident that his performance would be admirable. So I was unprepared for the call that very evening.

"Come pick up your male. He doesn't seem to understand what he is supposed to do."

I would have hoped that CC might have told him more of the birds and bees, and less about chasing sticks or bright red balls. But there is no question in my mind that CC told Copper many things, some of which stood him in good stead. She showed him how to swim, for instance, something she had learned quite naturally the very first time she'd approached Lake Onota. If CC enjoyed chasing sticks in the field, pursuing a stick thrown in the water was heaven itself. She swam with an assured grace, effortlessly, quietly, and straight as an arrow. Copper watched from the bank, hopping about and barking excitedly, but unable to take the necessary plunge.

One day I pushed him off the edge of a dock. He thrashed about in obvious fear for his life, his head held as high above the surface of the water as he could manage, his front paws pumping feverishly up and down as he made his way to the shore. But at least now he knew that he could stay afloat, and immediately took after the stick in competition with CC. He was no match for her in the water, splashing along with his exaggerated style—plump, plump, plump. It took Copper three years to learn how to perfect his swimming stroke, and even then the poor fellow never managed his mother's grace in the water. But it hardly kept him from playing the game. He'd swim along behind CC and as she turned for home with the prize in her mouth, he'd catch her coming in and grab one end of the stick, and they'd proceed to shore each claiming a share of the prize.

I am of the opinion that Copper was aware of his relationship with CC. He took advantage of her, as many of us do to forgiving mothers. He'd be the first to the food dish of his choice, or he'd push her out of the way if the door was opened for them to proceed outside. He never put up with anyone patting CC when he was about. Indeed, he'd push her out of the way so that he alone would gain the attention. She never seemed to resent his aggressiveness. And yet, there were times when she'd put him very much in his place, with a growl or a snap, if he was offending her in some way. But he seemed to know how far he could push her.

He was perfectly willing to join in the ball games in the back field, which CC would generally inspire, but only for a reasonable period of time. He proved faster of foot than his mother, and keener at discovering the tossed ball that neither had seen thrown. This, of course, was half the fun of the game, to sniff out the unseen ball, ranging back and forth until picking up the scent, and to return with the prize. If it was Copper who found it, he'd run back purposefully and sit directly in front of me, allowing me to remove the ball. But if it was CC who found the ball, Copper would harass her as she tried to make her way back, nipping at a front leg, or a hind leg, and barking with obvious glee and excitement. CC, rarely annoyed, would only occasionally resent Copper's insistent attack and deliver a warning growl without ever dropping the ball. For the most part, she'd work her way through Copper's harassment, trying to catch him off guard, finally making a desperate run for it. But for obvious reasons she was reluctant to release the ball, for fear that Copper would pick it up and I'd have to pry it from her jaws. CC never tired of playing ball. Copper would often retire from the game and watch as CC scurried out into the field—if, that is, we hadn't tired of the game already.

Each night the dogs proceeded with me to the kitchen for a cookie before bedtime. They both sat expectantly while I dug out two dog biscuits. Copper always snapped at the biscuit and one had to be fast to avoid his teeth. CC had a soft mouth and gracefully eased it from my hand. And then it was off to bed in the back hall, the self-same space that had held the whelping box in Copper's formative days. They each had an old rug, and each had a favorite spot in which to lie. Copper rolled over, belly up, inviting me each evening to rub his stomach. CC preferred to have her ears pulled and her back rubbed, and they expected such treatment each evening.

Copper had originally resented being closed up each night. One evening I had left a freshly cleaned blue suit hanging on a peg in the back hall and found it in the morning on the floor, the shoulder having been gnawed out of the left sleeve. In a matter of hours and after a bit of casual observation, it became obvious that Copper was the culprit. It was fortunate indeed that the store from which the suit had come could not only match the material, but was willing to reconstruct the shoulder, else Copper might have experienced a more stern reprimand than he actually received.

There were times when we'd make a visit to the vet's for a shot or some other necessary medication. Copper would be frightened to death, and CC calm and restrained. In fact, Copper would crawl up into my lap, all one hundred pounds of him, forgetting to leave a hind foot on the floor. He'd cry and whimper minutes before he was led into the back chamber for the shot. Once there he'd shake at the presence of the doctor, the same kind vet who had brought him into the world. And he was most appreciative if you'd commiserate: "Oh, poooor Copper!" He'd slide up to you whimpering and crying and bury his head in your lap, begging for more sympathy.

I am convinced that Copper loved the companionship of people more than of dogs. There were many times when a collection of dogs would be playing about the grounds and Copper would appear to be enjoying the games. But if he caught sight of anyone in the family, he'd abruptly leave the dog games and come wagging up to us, where he'd stay.

School Days

IT IS A WONDER THAT EITHER OF our dogs responded to discipline. But generally they did, although CC could on occasion be quite obstinate. As are many owners, we were moved to take the dogs to obedience school at the local YMCA. Copper was only about six months old, and CC was just under two years. These outings were occasions of great excitement for the dogs, being thrown into a room with so many other dogs, large, small, menacing, and cuddly. I was elected to be Copper's handler. My wife handled CC, who was restrained—almost aloof. Copper was afraid and could hardly contain himself. When the instructor demonstrated the moves and the handlers were gathered around the edge of the room with dogs on the leash, Copper would crawl up into my lap and whimper fearfully. When it came time to perform, he was dreadful. Still, in time he seemed to understand, and he tried so very hard to please me, though I never felt sure that he really understood. He never saw much sense in the command to "stay," and yet he could be controlled as long as you kept eye contact with him. Turn your back and he would be on you in a flash.

I suppose it was not unusual that we never followed through with the training. Most people don't, and then blame the dog for not having accomplished a full course of discipline in five or ten outings of one-hour's duration. Yet our dogs always did respond to discipline, even when in the presence of a pack. If they could see or hear me and if they sensed that I meant business, they responded immediately, albeit CC somewhat reluctantly. They would come at a command, even heel—not quite by the book, but good enough for purposes, for instance, of not wandering out into a road as we walked along the edge of a thoroughfare. They would even stay until, of course, you'd go out of sight; then they didn't figure it counted. And maybe it shouldn't.

There have been frequent occasions when we'd walk around the block, taking us by a house where the yard is enclosed by a wooden picket fence. Within the area of the fence lives a feisty little dog, a miniature schnauzer, who on each occasion of our visits would rush along behind the fence barking and posturing and threatening to devour us all as we proceeded by. Our dogs preferred to ignore the schnauzer and walked sedately by our side—much, I suppose, to the consternation of the defending dog. On one such trip, some work was being done in the yard in question, and I could see that a section of the fence was down toward the end. On this day the schnauzer was out and up to his usual routine, dashing along snapping and barking, until, that is, he ran out of fence. Suddenly the dogs were face to face with no restraining barrier. There was a moment of indecision and silence on the part of the schnauzer when he made this discovery. Our dogs then did the usual friendly sniffing, which the schnauzer quietly endured with something of a visible tremor. And when the amenities were completed, he simply turned his back and strode back toward his house without sound or backward look. So much for noisy fellows who do their worst barking behind safe barriers.

There were occasions when the dogs would be off "on

the town," and we'd worry about the kind of trouble that they might create. Generally this occurred in the summer, when it was so hot that the lake had an obvious fascination. Swimmers generally welcomed our kind of dogs, who offered no threat but only asked for a chance to fetch a stick. CC, once at the beach, would spend the day untiringly at the chase. It is the male who is supposed to be the wanderer among hunting dogs, but not in our family. It was always Copper who would tire of the game and wonder what was going on at home. He'd come back hours before CC. And when he did, soaking wet, I'd know that it was time to fetch CC. We have a leash law in town, one of man's irrational laws that are unenforceable because there are so many offenders that it is folly to try and restrain all who break them.

One hot day when CC and Copper were missing, I received a phone call from an anonymous caller.

"Your two dogs are at the beach and somebody has called the dog officers. You'd better get down and pick them up."

I raced to the beach. There were about five dogs all enjoying the swim, and three of them were goldens. I whistled. Copper stopped his game and looked around, spied me, and rushed to my side, wagging his tail with delight. I stuffed him in the car and went back for CC. Whistles meant little, so I walked to the edge of the water, where two goldens were vying for a stick that a young swimmer had tossed. Goldens are fluffy and soft when dry, but slick and thin when wet. I was confused. I wasn't sure which one was CC. I called her name and neither dog responded. I wanted to walk into the water and check more closely, but had my clothes on, so I asked a young man if he'd mind leading that dog in.

"Is that your dog?" he said.

"I think so," I responded. I shall never forget his retort: "Whatsamatter? Don'cha even know your own dawg?"

It was a particularly embarrassing moment, and possibly I shall never forgive CC for putting me in such a situation. She simply refused to acknowledge me because she did not want to give up her afternoon swim. Copper watched from the car, pleased to have us both come toward him, as I scolded CC along the way.

Copper the Fighter

COPPER WASN'T MUCH OF A fiGHTER. I don't suppose he felt the need, nor do I think he had the stomach for it. He was fortunate enough to have steady female companionship, which, for instance, made him the envy of his peers during a stay at the kennels. They were generally housed together on such occasions. And CC certainly didn't seek or welcome stray males, being a spayed bitch and by nature aloof. So Copper rarely had to display a protective posture for the privilege of attending his lady. There were a few instances when his brother Nugget was in residence during a visit from our daughter's family. Copper took exception to the amorous attentions of Nugget to CC, and was sometimes offended by Nugget's aggressive approach to a community bone. Because of their long association, we tended to view such squabbling as petty irritability. Only once do I recall Copper's manhood rising sufficiently to the surface to cause him to display bold aggression. It was in the summer, after a trip to our summer home in Maine, a place to which we had only recently introduced the dogs. They had never before experienced the surf of the ocean, and accepted it as a titillating bonus in the usual chase of a stick. On this particular day, some folks had intruded on what we had come to consider our own private stretch of beach. They had a large male golden who immediately fancied CC, although she kept her distance with the necessary snaps and snarls to discourage his attentions. Copper, apparently secure in his knowledge that CC could fend for herself, paid little attention to the intruder—until, that is, the big fellow approached and sank his teeth into Copper's shoulder. This was perhaps the visitor's way of displaying his manhood to the female. There ensued a violent scuffle, which I managed to break up with a handy piece of driftwood, fearlessly and foolishly striding in between the snapping jaws and the entangled forms of the struggling dogs. Once separated, the dogs eyed each other briefly, and then Copper, with an uncharacteristic snarl, darted toward the bigger adversary and managed to grab the dog on his shoulder, hanging on until I once again managed to separate the two. Truthfully, I almost hated to spoil Copper's advantage. I was that pleased at his display of courage. I did praise his actions as the visiting family packed up and retired from "our" beach. It was total victory for the good guys on the strength of Copper's macho display. He seemed pleased.

Copper's Last Days

IT WAS EARLY IN THE SPRING OF 1981 that my son approached me at my typewriter in my upstairs office and said, "What's happened to Copper? He's lying down in the drive, frothing at the mouth, and he can't seem to get up."

Poor Copper. We rolled him onto an old rug, hefted him into the back of the station wagon, and were off for the vet's in a matter of minutes.

The vet pulled back his chops and said, "His color is terrible. He looks as if he is bleeding internally."

Indeed, his belly was distended, and swung from his backbone, not unlike that of a pregnant cow. Only his eyes, which looked at us beseechingly, and a quiet thumping tail gave evidence that he was alive and aware of his plight.

The vet thought he might have been hit by a car, but there were no external signs of such an impact. We were ushered from the vet's operating rooms and in time advised that Copper had been "bled," seemed to be surviving, and would be retained at the hospital for continued surveillance. Daily reports indicated progress, to the point where Copper was eating again. On the fourth day, he was returned home with the advice that he be kept quiet. Normally an instruction to keep Copper quiet would be one of pure folly, but this time, the poor fellow, though obviously pleased to be home, was quite willing to remain quiet. He slept much of the day, gratefully acknowledging the loving attention that we gave him.

He was with us for four days, each day appearing stronger, and taking a greater interest in the life about him. The next morning I found him lying flat on his stomach, his four legs splayed outward. I felt his stomach. It was softer, and I knew the internal bleeding had started again. I rolled him over into a more comfortable position and called the vet, who said that he should be in his hospital in twenty minutes. So I went to the kitchen to have breakfast. As I sat there, suddenly in walked Copper. Slowly, with tail wagging, he approached me, and then collapsed at my feet. How he managed to get up and make his way to the kitchen I shall never know, but it was the last time he did. We made it to the vet's, but Copper of West Street was gone minutes later. He succumbed at the very conclusion of his tenth year. Poor Copper. An autopsy showed that he had been riddled with cancer.

A Frightening Fourth

I EXPECTED THAT CC WOULD MISS her son, who had spent his entire life as her faithful companion. I looked for signs of melancholy in her aging eyes. But instead I perceived a new independence emerging. She danced more, as she had in her youth. She asked for and received more attention. Could it be that she enjoyed her new role of being the only ant on the ant-hill? I don't suppose I shall ever know. But in the meantime we grew old together. We were both resting more than used to be our wont, and she grew thinner while I put on weight. But I don't think either of us was unhappy with our lot at the time.

This is a tale about the frailty and courage of age, which creep inevitably forward, manifesting themselves with each passing month. It is also about a dog's life, of which we knew something, but not as much as CC, the heroine of this tale. Of course she topped us all, being well into her fifteenth year, which they say translates into the equivalent of slightly over one hundred human years.

We spent most of the year at our home in Pittsfield, Massachusetts. But since the lovely summer manse in York Harbor, Maine, had been passed on to my wife upon the death of her mother, CC and her son Copper had spent a portion of their senior years with us on the Maine coast. We had thought it unfair to subject my wife's mother to the burden of these two creatures during our previous visits, a view that she obviously also shared.

The dogs took to the Maine surf with enthusiasm. They spent endless hours chasing a stick into the surf. Because our house was but a quarter mile from the broad and populated beach, it was a constant temptation, particularly for CC, to steal down to the water's edge and encourage someone to toss her a stick. On more than one occasion she had to be reprimanded, for dogs were not permitted on the beach. Copper generally preferred to spend the time at home with the family.

But on the day of this particular tale we had set out for our summer quarters on the third of July in a caravan of two cars. It was after Copper had left us; only CC remained. My wife rode with most of the family chattels, and I in the lead car with some clothes and with CC settled in the back. She had haughtily waited for me to heft her up into the station wagon, being no longer willing, or perhaps able, to leap up on her own. I had some concern that she might not ride as well as she had in the past; it had been a year since she had endured a long four-hour trip. But she didn't seem at all averse to entering into the experience, although I hardly believed that she was aware what she was letting herself in for at the time. Still, I suppose she figured that going along was preferable in every case to being left at home to "watch the house and grounds," her usual chore at a time of a departure.

It was hot and humid, and I rolled down the windows to catch the breeze that flowed through the car and ruffled CC's fur. Poor old lady with the fur coat who couldn't skinny down to her skivvies in the warm summer breeze. Yet I've heard that a dog's fur acts as insulation that tends to keep her body cool rather than increase the heat; logical, but not my theory necessarily.

CC spent much of the early part of the ride pacing from side to side in the back of the wagon, occasionally lying down to rest her legs. Sometimes she appeared to stare out the back window, and one might have assumed it was because she recognized my wife, or at least the car, which tracked our path throughout. It is hard to know what she was seeing those days, because our vet had said she was forming cataracts, a typical sign of age.

We had driven for two hours when I pulled up at a gas station, partially out of deference to CC's assumed needs, and partially because my kidneys suggested a stop to be in order. We gassed up, CC drank thirstily of some water, wolfed down a few dog biscuits, dutifully tinkled in some long grass nearby, and stretched her legs. Back in the wagon she settled down for a long nap, and didn't show her head until I started winding about on the crooked roads leading into York Village.

It was about two o'clock when we pulled into the drive, found the familiar old house looking splendidly yellow in its new coat of paint and the grounds nicely trimmed as per a contract with a gardener. CC seemed as interested and excited as my wife, who has been known to start her happy humming, as a little girl just out from school might, as she enters the hallowed grounds of her summer playground, which indeed it has been since 1926.

It was about two hours later, after we had unpacked and, along with CC, explored all nooks and crannies of the house, that a thunderstorm struck with vicious fury. Lightning flashed, thunder pealed, and water splashed. CC responded nervously, pacing the floor. Although she did not hear well, there was little question that she heard the rolling in the sky all too well. I went upstairs to shut windows and she followed at my heels, although she well knew that the upstairs had long been prohibited to ladies in fur coats. I did not press charges. She was frightened to death, and probably wondering why we had left home.

Once back downstairs, I settled down to read some old magazines I'd brought along just as a vicious clap of thunder followed the slashing brilliance of lightning that skipped over the ocean. It seemed but an arm's length away. CC popped up from an uneasy reclining position, squatted in the very center of the room, and left a perfect symmetrical pool of warm tinkle. My wife was off picking up the necessary groceries at the market, and fearful of her reaction, I roared my displeasure, ushered CC out the door to the covered porch, and rushed for the paper towels. CC stood shaking at the door and pleaded to get back in. I opened the door, and continued my mop-up job, which was so well accomplished that to this day the spot is not readily noticeable.

By nighttime we were somewhat settled. CC was not pleased to be incarcerated in the back kitchen for the night. She had never learned to feel at home in this particular area, but then, she had lately been reluctant to be put to bed even in her winter home, with which she was most familiar. I fancied that this was a problem of age, because it had not always been so. It was almost as if she had learned to fear the dark, that she didn't enjoy being alone and felt more comfortable with us nearby. I guess that most of us feel pretty much the same way as the years go by, though I know solitude can be a blessing on occasion. But it was the unusual night when CC did not wet the floor, and this particular night was hardly any exception. This nocturnal habit was definitely a product of advancing age, and in our advancing years we became adept at the cleanup.

The following day was the anniversary of the birth of the nation. It arrived clear and sunny, and was heralded by popping firecrackers. At noon, CC and I were whiling away the time on the porch when the neighbor two houses up the coast unleashed the first of his traditional series of aerial bombs. It shook the old house and almost blew me off my chair. CC snorted nervously and nestled her head in my lap. I took to watching the white puffs in the sky that heralded by seconds the crashing bang of the next bomb, and I tried to cover CC's ears with my hands. After the third bomb she tried to get into the house, but was denied access by my wife, mindful of the debacle of the previous day. So CC headed back toward the garage. We yelled at her to return. She stopped, looked back, and then proceeded, oddly enough, in the direction of the source of the disturbance. I incorrectly assumed that she was headed toward some favorite place that she knew as a temporary haven, and would return when the "shelling" ceased.

Ten minutes later I became disturbed. The bombs had been spent and CC had not returned. I followed the direction she had gone on foot, whistling and calling. I circled the block. No CC.

I asked the policeman who rested on the driveway gate of a neighbor, probably to ease the weight of the arsenal which hung from his belt, "Have you seen a golden retriever?" He shook his head. Poor CC. Where had she gone in this alien world full of thunder, lightning, and Fourth of July bombs?

I scoured the area unsuccessfully. We were due at the nearby club for lunch. I had to change my clothes, and as I did I comforted my wife, as well as myself, "She'll come home in good time."

But would she, in her harassed state? Being over one

hundred years old, and more to the point, being away from her home territory, I wondered if she'd even know or care where home was situated.

At the club we renewed acquaintances with some friends and I ordered lunch. As my lunch arrived, so too did the club manager. "Mr. Jackson, you had a phone call. Your dog is down at Lusty's and is very agitated. They are trying to hold her. Would you go down there?"

Of course! But Lusty's is a marina across the harbor, probably three to four miles away. It is on a peninsula that houses not only the marina and its boat yard, but also the Dockside Restaurant and a series of cottages surrounding a main house.

I left untouched the summer's first lobster roll, potato puffs, and ice tea, and dashed for home to pick up a leash and the car. I inquired at the marina for the golden. No one knew about her. I asked at Dockside. They shook their heads. "Try," they said, "Dave Lusty's guest house." I did.

There was CC wet and forlorn, at the end of a long blue-and-white line, tethered to the bannister of the porch steps. She was shaking, was soaking wet, looked thin and frightened. She saw me approach, and seemed immediately relieved. I slipped on her leash and comforted her.

We went up the stairs to the room marked "Crew's Quarters." Six people in unison said, "You're Mr. Jackson? Oh, thank God. Your poor dog has been frightened to death." CC looked up at me and wagged her tail. But there was little joy left in her.

"Where did you find her?" I asked.

Everyone talked at once. "Some of our guests saw her go into the water across by the condominiums."

"You mean she crossed the channel?" I asked incredulously. That, I knew, was the swiftest water about. There is a narrow right-angle turn in the channel, the throat to the harbor, where the water is squeezed and courses through at the whim of a seventeen-foot tide, and is either carried out to sea or into the harbor and up the York River. Evidence of its swiftness resides in the black buoy that bends in the tide's direction and ceaselessly tugs at its sturdy mooring, a monument to man's defiance of the will of nature. Poor CC. Why did she chance it? How did she make it?

"And the sad thing was," said Dave Lusty, "some folks right here were setting off bombs just as she reached this shore."

"But how could she make it? She's an old girl. She must have been very upset," I said.

"They're smarter than you think," rejoined Lusty. "We used to have a golden, and she'd swim that channel. She'd seem to gauge the tide and start off accordingly, so that as she was swept along she would end up about where she wanted to go."

"At fifteen years of age?"

"No...."

Poor CC. She'd never attempted that swim before. How could she know how it should be done?

I made my thanks to all the "crew" and CC and I went to the car, where she patiently waited for me to boost her into the back. My wife was still at lunch on the terrace. There was a three-piece band playing old favorites, and people down below on the beach were taking advantage of the music and dancing on the sand. Despite her grueling experience, CC was not admitted to the terrace. It's out of bounds to dogs. So I moved my lunch to an end table where we could communicate by touching one another as she lay off the terrace.

That evening as night's curtain descended, the club's

annual fireworks display rose up from the beach at precisely nine P.M. I was ready, even if CC wasn't. We spent the time together deep in the laundry room of the oversized cellar. You could hear the rockets and bombs, but they seemed somewhat removed.

I guess one might conclude that it is not all beer and skittles—this getting old. I don't know why we should get more timid with the passing years, with all that experience we've packed into the years we've managed to survive. As time went on CC developed a lump the size of a tennis ball on her right flank and another cyst on the top of her spine, and you could count the number of vertebrae on her back. She stumbled sometimes and she had grayed to look more like a raccoon than a dog. She did wag her tail, the same tail that she had perversely chewed toward her last years so that the handsome "feathers" were gone and it looked more as a ragged rat's tail might. But she was still a handsome, dignified creature until she got so weak that she couldn't or wouldn't eat. My wife and I made a decision. We went together to see our friendly vet, and with her head in my lap, he administered the shot that brought her peace.